Mehrez Ben Nasr

Study of the impact of factors on returns on financial assets

Mehrez Ben Nasr

Study of the impact of factors on returns on financial assets

ScienciaScripts

Imprint

Any brand names and product names mentioned in this book are subject to trademark, brand or patent protection and are trademarks or registered trademarks of their respective holders. The use of brand names, product names, common names, trade names, product descriptions etc. even without a particular marking in this work is in no way to be construed to mean that such names may be regarded as unrestricted in respect of trademark and brand protection legislation and could thus be used by anyone.

Cover image: www.ingimage.com

This book is a translation from the original published under ISBN 978-620-6-71507-8.

Publisher:
Sciencia Scripts
is a trademark of
Dodo Books Indian Ocean Ltd. and OmniScriptum S.R.L publishing group

120 High Road, East Finchley, London, N2 9ED, United Kingdom
Str. Armeneasca 28/1, office 1, Chisinau MD-2012, Republic of Moldova, Europe
Printed at: see last page
ISBN: 978-620-7-98692-7

ACKNOWLEDGEMENTS

This work was carried out under the scientific supervision of Mr /avier Bouna Niang, Managing Director Typhoon Partners, London UK and Ms Nozha Karmous, Quantitative Analyst HSBC, Paris France. I would like to thank them most sincerely and respectfully for having directed this work, and for their constant advice and encouragement. Their good humour, patience and kindness were always able to soften even the most tense moments. I would like to express my deepest gratitude.

To Mr. Hichem Rammeh, my university tutor and head of the Actuarial Master's program at Paris Dauphine University in Tunis, I am very grateful for all the help he has given me, and for his modesty and understanding. Please find here the expression of my deep respect and gratitude.

I sincerely thank all the members of my jury for agreeing to judge my work.

I would also like to extend my warmest thanks to Mrs Claudine DHUIN, MIDO's coordinator with Dauphine Tunis, for her availability, help, encouragement and invaluable advice.

I would also like to thank all my colleagues and classmates for their sympathy and moral support, and for the good-humored atmosphere they maintained.

I would also like to thank all those who contributed in any way to the realization of this work.

Summary

This dissertation is an internship report for Typhoon Partner, a UK-based proprietary investment firm specializing in the development of quantitative investment strategies.

We will focus on intra-sector and inter-sector returns relative to the fundamental characteristics of US equities. After a thorough analysis of the basic concepts of risk and the issues related to risk management, we will first look at the various risk measures associated with each sector (VaR, Expected Shortfall, etc.). Secondly, using the relevant literature, we apply a statistical model to explain cross-asset returns in the equity universe using the financial ratios selected, with the aim of studying the performance of equities within a sector and between several sectors. Finally, we look at the comparison of investment strategies and portfolio insurance.

Keywords: Equity, SPDR ETF, FM Cross-sectional regression, Value at Risk, Expected Shortfall , Investment strategy, Portfolio insurance.

Abstract

The present work is a report of the internship realized in "Typhoon Partner", an investment company for own account domiciled in the UK which is specialized in the development of quantitative investment strategies. We will focus our study on intra-sectoral and inter-sectoral returns relating to fundamental characteristics of US actions. After having analysed the basic concepts of risk and issues related to the management of these risks, we are interested in the first time to the various measures of risk associated with each sector (VaR, Expected Shortfall, ...). Second, we apply, with the relevant bibliographic data, a statistical model explaining the cross-asset returns of the universe of equities using financial ratios used in order to study the actions of performance within a sector and between several sectors Finally, we will focus on the comparison of some investment strategies and portfolio insurance.

Keywords: Stocks, SPDR ETFs, FM cross-sectional regression, Value at Risk, Expected Shortfall, Investment strategy, Portfolio Insurance.

TABLE OF CONTENTS

INTRODUCTION

The world has experienced a number of banking and financial crises, the most recent of which began in the United States in autumn 2008, before spreading to every country in the world. As a result, risk analysis and assessment, which involves risk analysis, the study of risk indicators and the calculation of risk measures, is now a very active area of study. Financial risk management is the immediate solution to which actuaries, economists and financial managers have turned since the last crisis, with the aim of better studying the financial market in order to minimize the potential losses of the assets that make up a portfolio.

As a result, the scope of actuarial activities is no longer limited to insurance companies, but has expanded into the field of finance. The considerable development of financial markets is linked to new risks. This was a strong incentive for me to do my internship with an investment fund.

Our current project is part of this effort to improve current risk indicators. Our objective is to study intra-sector and inter-sector returns relative to the fundamental characteristics of US equities. To this end, we are interested in the practical implementation of a statistical model explaining cross-asset returns in the equity universe. Our study is divided into five chapters. The first chapter introduces the basic concepts of risk and the issues involved in risk management. After defining and identifying the various risks, we define the financial instruments that will be useful to us and that enable us to hedge these risks. The second chapter will be devoted to the financial theory studied during the internship, and to the definitions of the variables we are interested in in this work. The third chapter covers the various mathematical and statistical tools used during the internship. The fourth chapter is devoted to defining the methodology of the work, the experimental part and the analysis and exploitation of the data. Finally, the extension to investment strategies and portfolio insurance will be the subject of the last chapter.

We close our report with a general conclusion, outlining the main results obtained.

Chapter I Theory of risk

In this chapter, we introduce the basic concepts of risk and the issues involved in managing these risks. After defining and identifying the various risks, we define the financial instruments that will be useful in our work and that enable these risks to be hedged.

1.1 Introduction

A risk is a potential event whose occurrence is likely to have a negative impact on a person or organization. Risk can be defined as the variability or volatility of an unforeseen outcome. It is generally measured by the variance or standard deviation of past results, which corresponds to the historical approach to risk measurement. While all businesses are exposed to uncertainty, financial institutions face special types of risk due to the specific nature of their activities.

The general objective of financial institutions is to maximize profit and shareholder value by offering a variety of financial services while optimizing the risks incurred.

There are several ways of classifying risks. The first is to distinguish between business risk and financial risk. Business risk is linked to the company's own activities. This risk concerns factors affecting the product (or market).

Financial risk relates to potential losses on the financial market caused by movements in financial variables (Jarion and Khoury 1996, p2). It is often associated with leverage, leading to the risk that liabilities and debts will not match current assets (Geleason 2000, p21).

Risk can also be broken down into systematic and non-systematic risk.

Systematic risk: Equity risk can be general or company-specific. General (or systematic) risk refers to price fluctuations resulting from general stock market trends. For example, the value of a share may fall on the markets without any change in the company's economic situation.

Unsystematic risk: Specific or unsystematic risk depends on the company concerned. For example, certain factors can depress a share price even though the stock market or the prices of similar companies are rising. Specific risk factors include negative events (strikes, management crises, poor annual results) as well as good news (the signing of a major contract, innovative products, good sales prospects). Random events within a company have an impact on share price fluctuations (volatility), but cannot be anticipated.

Whereas systematic risk is associated with the market or the state of the economy in general, unsystematic risk is linked to a specific company. While unsystematic risk can be mitigated by portfolio diversification, systematic risk cannot be eliminated by diversification. Portions of systematic risk can be reduced through risk mitigation and risk transfer techniques.

Techniques used to protect against risk include standardizing all activities, diversifying the portfolio and implementing an accountability and motivation plan.

However, there are some risks that can neither be eliminated nor transferred, and must be addressed by the institution. The first is due to the complexity of the risk and the difficulty of the asset with which it is associated. The second risk is accepted by the financial institution because it is intimately linked to the institution (or it may fall within the institution's risk appetite). Examples of such risks include credit risk, interest-rate risk and foreign-exchange risk.

1.2 financial risks

A financial risk is the risk of losing money following a financial transaction (on a financial asset) or an economic transaction with a financial impact (for example, a sale on credit or in foreign currency).

The main types of financial risk are as follows:

- **Market risk**: is the risk of loss that may result from fluctuations in the prices of the financial instruments making up a portfolio. The risk may relate to share prices, interest rates, exchange rates, commodity prices, etc.... This risk is generally measured by market volatility, a statistical figure which cannot, however, fully reflect all the uncertainties inherent in the markets, let alone the economy in general. Market risk is expressed by the risk premium for the market in general, and by the beta coefficient for the evolution of the price of a particular asset in relation to the market.

- **Counterparty risk**: counterparty risk corresponds to the credit risk arising from the caisse's current and potential over-the-counter financial instrument exposures.

Derivative financial instrument transactions are carried out with financial institutions whose credit ratings are established by recognized rating agencies, and whose operating limits are set by management. In addition, the caisse enters into legal agreements based on International Swaps and Derivatives Association (ISDA) standards, enabling it to benefit from the offsetting effect between the amounts at risk and the exchange of collateral, in order to limit its net exposure to this credit risk.

- **Credit risk**: represents the possibility of incurring a loss in market value in the event that a borrower, endorser, guarantor or counterparty defaults on a loan or other financial commitment, or suffers a deterioration in its financial situation.

Credit risk analysis includes measuring the probability of default and the recovery rate on debt securities held by the credit union, as well as monitoring changes in the credit quality of issuers and groups of issuers whose securities are held in all of the credit union's portfolios. As part of its credit risk management, the caisse frequently monitors changes in credit ratings from rating agencies, and compares them with internal credit ratings where available. It also uses credit VaR for its specialized bond, long-bond, real-return bond and real-estate debt portfolios.

Credit risk and counterparty risk are two notions that are sometimes interchangeable. For example, let's say you want to buy a Spanish Treasury bond, and you also want to hedge against the risk of Spain defaulting. We therefore buy a CDS (Credit Default Swap), a contract that acts as insurance against a possible default.

The Spanish bond can be described as a counterparty risk as well as a credit risk: this corresponds to the fact that Spain may default and not repay us all the funds we have lent it (as in the case of Greece in 2008).

We can only talk about counterparty risk on the CDS: the bank that sold us this CDS is supposed to, in the event of Spain defaulting.

- **Liquidity risk**: this is the risk that the Group will be unable to meet its financial liabilities on an ongoing basis, without having to r a i s e funds at abnormally high prices or sell assets by force. It also corresponds to the risk that it will not be possible to disinvest rapidly or to invest without having a marked and unfavorable effect on the price of the investment in question.

- **<u>Interest-rate risk</u>:** structural (as opposed to market-rate risk): this is the risk associated with loans, i.e. generated by a mismatch between assets and liabilities. It is the risk that credit rates will evolve unfavorably. For example, a variable-rate borrower incurs interest-rate risk when rates rise, because he has to pay more. Conversely, a lender is at risk when rates fall, as it loses income.

For a bank, this is the risk that changes in market rates will lead to a higher cost of remunerating deposits than the gains generated by interest on loans granted.

- **<u>Weather risk</u>:** this is the risk of potential loss of sales or profit due to variations in the weather. It concerns the four main weather families: temperature, precipitation, sunshine and wind. Weather risk concerns only ordinary variations in the weather. It concerns the potential impact on a company's performance of a weather anomaly, i.e. a fluctuation around its average value. In meteorology, the average (also called the normal) is generally calculated over 30 years. It should be noted that this risk is not significant for banks, except perhaps for commodities.

- **<u>Country risk</u>:** strictly speaking, country risk is the probability of a country defaulting on its foreign debt. On the other hand, if a country experiences a very serious crisis (war, revolution, cascading bankruptcies, etc.), even "trusted" companies, despite their credibility, will find themselves in difficulty. This is a counterparty risk linked to the counterparty's environment.

- **<u>Operational risk</u>:** for financial institutions (banking and insurance), operational risk is the risk of direct or indirect loss due to inadequacy or failure of the institution's procedures (absent or incomplete analysis or control, unsecured procedure, fraud), its staff (error, malice and fraud), internal systems (IT failure, etc.) or external risks (flood, fire, etc.).

- **<u>Concentration risk</u>:** the risk associated with a high concentration of investments in certain asset classes or markets. The higher the distribution within the fund, the lower the concentration risk.

- **Basis risk:** linked to the evolution of an underlying price in relation to that of its hedge (put, future contract, etc.). As the hedge is not always perfectly adapted, a price differential can arise, known as basis risk.

- **Inflation risk** is the risk associated with inflation. In concrete terms, as inflation rises in a given country, the purchasing power of that country's currency declines. Bonds guarantee a specific nominal rate. By deducting inflation from this nominal rate, we obtain the "real rate". Consequently, the higher the inflation rate, the lower the real rate (which translates into a fall in the value of the bond).

- **Performance risk:** Risks that affect performance. This risk is a combination of market risk and the manager's active policy. Risk may vary according to the choices made by each UCI and the existence, absence or limitations of any third-party guarantees.

1.3 Definition of financial instruments

In this section, we introduce the various financial tools that will be used in this work.

1.3.1 Contracts

A futures contract is a transaction negotiated between two counterparties (the buyer and the seller) on an organized, regulated market known as a "futures market". It represents a commitment to buy (for the buyer) and sell (for the seller) a product at a predetermined price and date. Futures contracts are used for both hedging and speculative purposes. It is not an option. Buying or selling is mandatory at maturity.

The components of a futures contract are as follows:

- **The underlying:** this is the asset on which the contract is based. The underlying asset of a futures contract can be a physical asset (commodities or agricultural products), a financial instrument (shares, bonds, interest rates, exchange rates) or a stock market or weather index.

- **Term:** this is the date on which the contract expires. For futures contracts, expiry dates are standardized (third Friday of the expiry month).

- **The price:** this is the amount at which you agree to buy or sell the underlying asset.

Futures market players :

- **Speculators**: these speculate on the value of the goods linked to the futures contract, taking a higher-than-average risk in order to achieve greater profit potential.

- **Hedgers**: the aim of hedgers is to use a futures contract to protect themselves, i.e. to hedge against possible volatility in the futures contract item, they are direct users of the hedged item.

Example: a forward contract is an agreement between two parties to buy or sell an asset at a future date, for a price fixed in advance. To find out how much a forward contract is worth, all you need to know is the relationship between the forward price and the spot price:

Simple forward price. Let r be the risk-free interest rate, in this case :

$$F0 = s_0 \exp(rT)$$

Forward price with dividend rate. If the underlying pays dividends at the rate q (for example, a share or a Price Return index), the equation becomes :

$$F0 = s_0 \exp((r-q)T)$$

1.3.2 ETF (Exchange Traded Funds) or trackers

An Index Tracker, or Exchange Traded Fund (ETF), is a type of securities investment fund that replicates a stock market index (unfunded index), an asset or simply an exchange-traded strategy.

These funds are managed in France by undertakings for collective investment in transferable securities, also known as index funds.

Tracker principle :

The purpose of an index tracker is to replicate the performance of an equity, bond or commodity index. Most trackers replicate a general stock market index or a sector index, such as companies in the pharmaceuticals sector, regardless of the stock market on which each is listed.

Unlike conventional investment funds, trackers do not require the assistance of financial analysts. It is therefore a passive management tool. Its supporters believe that financial analysts are not capable of beating stock market indices over the long term, so there's no point in paying for their services, and that it's better to buy a tracker that replicates a stock market index.

In France, trackers are divided into four categories:

- Market index trackers.
- Trackers on commodity indices.
- Strategy index trackers.
- Active trackers: unlike the other categories, active trackers don't just track their benchmark index. There are leveraged funds, capital-protected funds and funds that outperform the index.

Advantages of trackers:

- **The simplest way to invest in the performance of an index**: trackers are the simplest way to invest in the performance of an index, since by buying a single security you can perfectly replicate the index's performance.
- **Transparency**: You can find out the tracker's value at any time, every 15 seconds. The indicative net asset value (iNAV) update enables instant comparison of a tracker's market price with its theoretical value.
- **Reducing risk through diversification**: trackers give investors access to a diversified portfolio of equities in a single transaction, at a reduced price.
- **Liquidity:** Trackers are continuously listed and trade in the same way as ordinary shares. They can be bought and sold through their usual intermediaries.
- **The subscription and redemption process:** one of the key features of trackers is the in-kind subscription and redemption process. Through subscription, an investor can acquire a certain number of trackers (subscription unit) in return for delivery of the

underlying basket of shares plus a balancing payment. Redemption enables the investor who acquires a certain number of trackers (redemption unit) to receive the underlying basket of shares in return.

Disadvantages of trackers:

- **There is a risk of default by the bank:** the synthetic tracker is a portfolio of shares that have nothing to do with the given index, but whose performance is similar. Its manager has signed a swap contract with an investment bank to guarantee the fund's performance. This reduces costs, but there is a risk of default by the bank.
- **Very high volatility:** Some index funds use leverage, a stock market mechanism that multiplies gains by buying stocks with money you don't own. They can double or triple your investment. However, trackers are highly volatile, since they amplify the slightest variation in the index they track.

1.3.3 Performance analysis of a portfolio

One of the aims of performance analysis is to compare different managers. In 1993, the Association for Investment Management and Research (AIMR, USA) defined standards which have since been revised. At global level, these are now known as GIPS (Global International Performance Standards).

Asset profitability :

$Pt-1$ is the price of an asset at time t and Dt the dividend paid between t-1 and t included. The profitability of asset Rt at time t is given by :

$$Rt = \frac{Pt + Dt\ Pt\text{-}1}{\text{-}1}$$

Logarithmic profitability is defined by $r\,t = \log (Rt + 1)$.

Note that the first formula is a D0LL to order 1 of logarithmic profitability.

Profitability between t-k and t :

$$Rt\text{-}k,t = (1 + Rt\text{-}k+1)\ (1 + Rt\text{-}k+2)\ldots (1 + Rt) - 1$$

Logarithmic profitability between t - k and t :

Empirical profitability :

$$r_{t-k,t} = \sum s = t-k+1^t \; r_s.$$

Arithmetic mean :

$$R_a = \frac{1}{T} \sum_{i=1}^{T} R_i.$$

Geometric mean

$$R_g = \left(\prod_{i=1}^{T} (1 + R_i) \right)^{1/T} - 1.$$

Empirical portfolio risk indicators :

Knowing a portfolio's return is not enough to assess its quality: two portfolios can have the same average return, but not the same behavior. In particular, they may fluctuate more or less around their average return, making them more or less unpredictable, or more or less risky in the sense that they may take on very unfavorable values if they deviate too far from their average trend.

The most common risk indicator is the (empirical) variance:

$$Var(V) := \frac{1}{T} \sum_{t=1}^{T} \left((R_t - R_m) \right)^2$$

$$R_m = \frac{1}{T} \sum_{t=1}^{T} R_t$$

Historical volatility :

Volatility is the extent to which the price of a financial asset varies. It serves as a parameter for quantifying the return and price risk of a financial asset. When volatility is high, the possibility of gain is greater, but so is the risk of loss. This is the case, for example, with the shares of a company with more debt, or with greater growth potential, and therefore a higher-than-average share price. If sales growth is slower than expected, or the company struggles to repay its debt, the share price will fall sharply.

The method used to calculate this approximation of risk is, however, contested, since it assumes that future trends will be inspired by past trends. This quantification uses the standard deviation of historical variations in profitability. In other words, another simplification: it is based on the more or less Gaussian curve of past price rises and falls for this asset, over a series of historical periods. For example, to take extremes, it uses the standard deviation of daily variations over a month, or monthly variations over ten years, etc.

Implied volatility :

Implied volatility represents the volatility of the option's asset price returns, calculated iteratively. All the parameters characterizing the option are known: the strike price, the price of the underlying asset, the time to maturity, the risk-free interest rate, the dividend and the option premium observed on the market. This is a useful indicator for the investor, as it corresponds to the volatility anticipated by market participants for the life of the option, and is reflected in the option premium.

Note that the higher the implied volatility, the higher the option premium, and vice versa. Investors can thus compare implied volatility with the historical volatility of the underlying asset, and form their own opinion of future volatility to help them implement the option strategy they deem optimal.

Implied volatility is calculated from the price of options, which are used to bet or hedge against an extreme scenario. For example, a grain farmer who fears that the price of wheat will fall from 100 to 60 euros per tonne due to competition from new producing countries, will hedge with a put option at 70 euros per tonne. Should the price fall to 60 or 50 euros, the option gives him the right to sell at 70 euros, thus limiting his loss to 30 euros per tonne.

Chapter II
The financial theory proposed in the research articles studied

This section will be devoted to the financial theory studied during the course, as well as to the
definitions of the variables we'll be looking at in this work.

2.1 Motivation

The asset pricing theory attributed to Sharpe [1964] and Linter [1965] is an empirical failure. Beta is not sufficient to explain the cross-asset returns of the equity universe (Fama et en français [1992].) the cross-section of realized equity returns (Fama et en français [1992].) In contrast, variables unrelated to existing theory such as Market Value of Equity (ME) (Banz [1981] Book-to-market ratio (BE / ME) (Rosenberg, Reid and Lanstein [1985]), cash flow-to-price (C / P) (Lakonishok, Shleifer and Vishny [1994]), and past returns (Bondt and Thaler [1985], Jegadeesh and Titman [1993], Moskowitz and Grinblatt [1999], and Asness [1997]) have significant explanatory power.

In the articles studied during this internship (see bibliography section), the authors tested whether a better prediction of the future returns contained in each variable is obtained by breaking the variable into two industry-related components:

The first component represents the difference between companies' overall market characteristics (e.g. their BE / ME ratios) and the average characteristics of their specific sector (e.g. the average BE / ME ratios of a given industrial sector) to obtain an within-industry variable (Goodman and Peavy [1983]).

The second component represents the average characteristic of the company's sector, known as the across-industry variable.

There can be several advantages to decomposition within t h e industry and in industry :

• Measuring variables relative to their industry averages can reduce measurement errors. For example, differences in accounting practices between industries can lead to differences in a variable that are unrelated to future returns.

- A company's risk and the probability of earning related economic rents may be more a function of the company's position within its sector than its position relative to all companies in the listed economy (Bain [1951], Collins and Preston [1969]).
- Portfolios formed by sorting stocks within their respective industry variables are more diversified in terms of industry representation than portfolios formed by sorting stocks on whole market variables.

2.2 Some definitions of the variables used in the bibliographical section

In the article studied, the authors explained the profitability of equities according to different variables. In what follows, we give the different definitions of these variables:

Market Value of Equity (ME):

The market value of equity is the total market value in dollars of all a company's outstanding shares. The market value of equity is calculated by multiplying the company's share price by its number of shares outstanding. The market value of a company's equity is therefore always changing, as these two variables change. The market value of a company's equity is different from its book value of equity, because the market value of equity does not consider the company's growth potential.

Book-to-market ratio (BE/ME):

The book-to-market ratio is a ratio used to find the value of a company by comparing its book value to its market value.

Book value is calculated by looking at the company's historical cost. Market value is determined on the stock market by reference to market capitalization.

Formula:

$$\text{Book -to- market ratio} = \frac{comptable\ value\ of\ l'entreprise}{\square\square l\square\square\square\square\square\ \square\square\square\square h\acute{e}\ de\ l'entreprice}$$

The book-to-market ratio attempts to identify under- or overvalued securities by taking the book value and dividing it by the market value.

In basic terms, if the ratio is greater than 1, then the stock is undervalued; if it is less than 1, the stock is overvalued.

Price-to-Book value :

The price-to-book ratio (P / B Ratio) is a ratio used to compare the market value of a stock with its book value. It is calculated by dividing the stock's current closing price by the last quarter's book value per share.

$$\text{P / B Ratio} = \frac{prix\ de\ l'action}{Actif\text{-}immobilication\ incorporelle\ et\ paccif}$$

A decrease in the P/B ratio could mean that the stock is undervalued. However, it could also mean that something is fundamentally wrong with the company.

The P/B ratio reflects the value that market participants attach to a company's capital stock in relation to its book value of equity. The market value of a share is a forward-looking measure that reflects a company's cash flows The book value of equity is an accounting measure that is based on the principle of historical cost, and reflects past equity issues, plus a n y profits or losses, and minus dividends and share buybacks.

ΔEMP :

A new measure of distress: percentage change in employees over the past year.

PAST (x,y) which is the monthly return over the last y months, excluding the previous (x-1) months.

Free cash flow per share:

Free cash flow per share is a measure of a company's financial flexibility, determined by dividing free cash flow by the total number of shares outstanding. This measure serves as a proxy for measuring changes in earnings per share.

$$\text{Free Cash Flow Per Share} = \frac{Free\ Cach\ Flow}{\#\ Sharec\ Out\ ctanding}$$

Chapter III
Statistical and mathematical tools

In this section, we introduce the various mathematical and statistical tools used during the course.

3.1 The regression model

This model was introduced by Sir Francis Galton following a study of the height of the descendants of tall people, which decreases from generation to generation towards an average height.

Regression is a set of statistical methods widely used to analyze the relationship of one variable to one or more others. In our study, we are interested in the two best-known regression models: linear regression and non-linear regression.

3.1.1 Linear regression Definition :

A linear regression model is a regression model of an explained variable on one or more explanatory variables, in which it is assumed that the function linking the explanatory variables to the explained variable is linear in its parameters. A linear regression model is one in which the conditional expectation of y knowing x is an affine transformation of x.

This model is estimated by several methods, such as maximum likelihood, least squares or Bayesian inference.

Theory behind :

The linear regression model is used to predict and explain a phenomenon.

After estimating a linear regression model, we can predict what the level of y would be for values of x, so this model allows us to estimate the effect of one or more variables on another, controlling for a set of factors.

The linear regression method can be considered as a method of
used to predict a quantitative variable.

Advantages and limitations of the linear regression model :

- Linear regression uses a statistical model which, when the relationship between the independent variables and the dependent variable is almost linear, gives optimal results.
- Linear regression is often misused to model non-linear relationships.
- Linear regression is limited to predicting numerical output. Figure 1 is an example of simple linear regression:

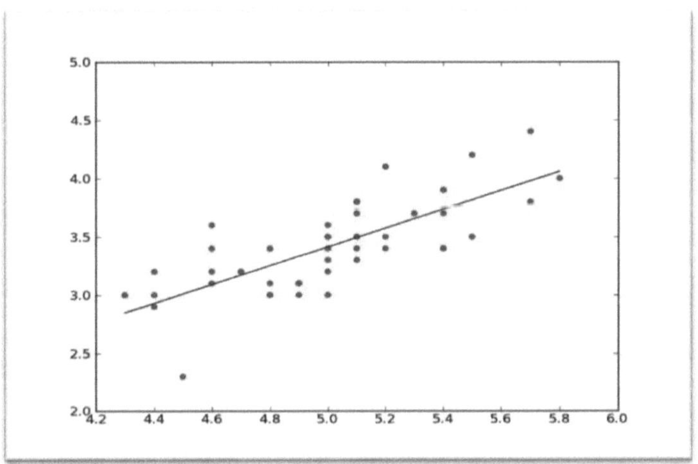

Fig. 1 Example of simple linear regression.

http://www.science-emergence.com/SimpleLinearRegressionPython/

3.1.2   Non-linear regression

The aim of non-linear regression is to fit a non-linear model to a set of values in order to determine the curve that most closely resembles the data curve of Y as a function of x.

Definition:

The non-linear regression model is written: $T_i = f(x_i, \Theta) + \varepsilon_i$

i=1....n,

- The probability distribution on ε_i is a normal distribution, reduced centered and of finite variance σ_i.

- The εi are independent of each other.

- The variable Yi represents observation i of the dependent variable.

- Θ represents a p-component vector of generally unknown parameters.

- The f function is the regression function, mostly non-linear. It depends on a real variable x and parameters Θ.

Figure 2 is an example of nonlinear regression with uncertainty bars:

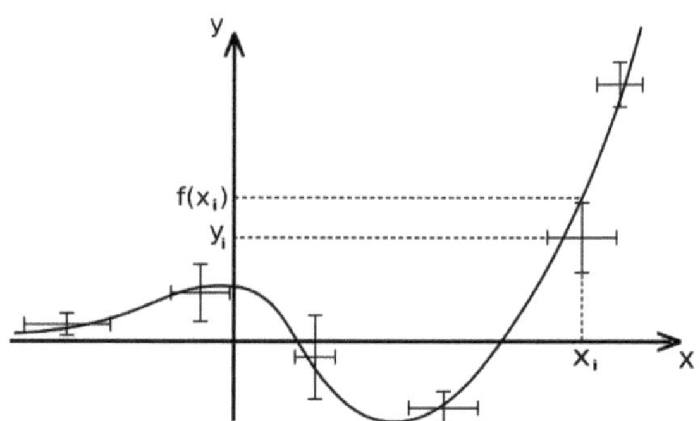

Fig. 2: Example of non-linear regression.

https://en.wikipedia.org/wiki/R%C3%A9gression_non_lin%C3%A9aire

Theory behind :

The aim of non-linear regression is to fit a non-linear model to a set of values in order to determine the curve that most closely resembles the data curve of Y as a function of x.

Advantages and limitations of the non-linear regression model :

- The greatest advantage of nonlinear least squares regression over many other techniques is the wide range of functions that can be shaped.

- Constraints can be easily integrated into a non-linear model, and are more difficult to apply to linear models.

- The parameters of a nonlinear model generally have a direct interpretation in terms of the process under study.

- One of the disadvantages of non-linear models is that the process is iterative. To estimate the model parameters, you start with a set of starting values supplied by the user.

- The non-linear regression model of the data is slightly more complicated adjustment of a linear model.

3.2 The factorial model

In 1976, Ross initiated the development of a model of the financial market over an extended period.
of time [0, T] where agents explain returns by the same factors. The factor model assumes that the rate of return on an asset is given by :

$$R_i = E(R_i) + b_1 f_1 + \cdots + b_k f_k + \varepsilon_i \qquad i = 1 \ldots n$$

where f_j $j = 1..n$ are random variables called factors, b_j are constants and ε_i is specific risk.

With $E(f_j) = 0$ for all j, $E(\varepsilon_i) = 0$, $cov(\varepsilon_i, f_j) = 0$, $cov(\varepsilon_i, \varepsilon_j) = 0$ if i G j.
F a c t o r models can be classified into three types: macroeconomic, fundamental or statistical.

Macroeconomic factor models :

These models explain asset returns through observable macroeconomic variables such as the inflation rate, the growth rate of the economy or of certain sectors of the economy, aggregate consumption, the unemployment rate... etc.

Fundamental factor models :

The factors are fundamental in the sense that they are supposed to describe asset properties such as the capitalization size of the issuing company, the dividend yield, sector classifications, the book-to-market ratio, etc. The model we will be studying and testing during this internship can be classified in this second category. The model we will be studying and testing during this internship can be classified in this second category.

Statistical factor models :

In this type of model, statistical methods such as principal component analysis are used to estimate factors and sensitivities. Factors are treated as unobservable variables. This type of approach can result in factors that can be interpreted as stock market indices as a linear combination of different starting factors. However, by increasing the type of starting factors (e.g. by mixing inflation, equities, GDP etc.) the final factors can become difficult to interpret and non-intuitive.

3.3 Measures risks

Let X be a future financial position representing the total value of a financial institution's investments at a date T. The objective is to determine a (deterministic) quantity $\rho(X)$ such that $\rho(X) + X$ is admissible, i.e. considered acceptable.

Definitions

Artzner, Delbaen, Eber and Heath have proposed four properties that should be possessed by a good risk measure (consistent risk measure, Artzner et al. 1997). Let X and Y be two market portfolios and R a consistent risk measure, we have :

- **Monotone**: If $X \geq T$, p.s., then $R(X) \leq R(T)$.
- **Positive Homogeneous:** $R(\alpha X) = \alpha R(X)$, whatever $\alpha > 0$.
- **Invariance by translation:** $R(X + \alpha) = R(X) - \alpha$, whatever the real α.
- **Sub-additivity:** $R(X + T) \leq R(X) + R(T)$.

The fourth criterion is the most important, because it represents the effect of portfolio diversification: a company that covers two risks requires no more capital than the sum of those obtained for two separate companies carrying these two risks respectively.

3.3.1   Value at Risk (VaR)

What is "Value at Risk"?

A widely used risk indicator since the 90s, Value at Risk (VaR) has become a benchmark risk measure. It is often used by insurance companies, major banks and asset management companies in the context of the new Solvency II and Basel II standards.

Figure 3 shows an example of Value at Risk 10% for a portfolio following a normal distribution:

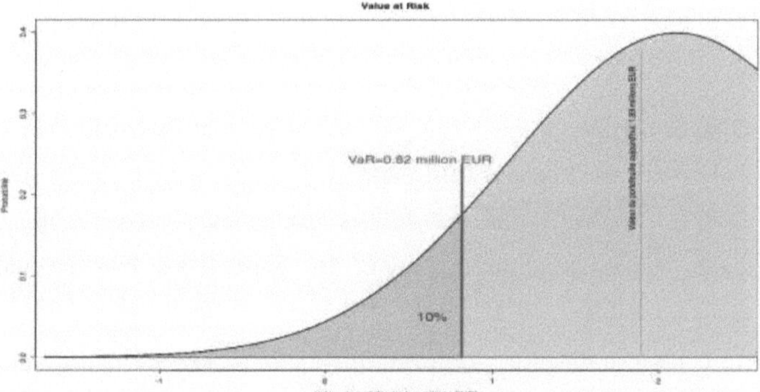

Fig. 3. 10% Value at Risk of a portfolio following a normal distribution.

https://fr.wikipedia.org/wiki/Value_at_risk

In its most general form, Value at Risk measures the potential loss in value of a risky asset or portfolio over a defined period for a given confidence interval. Thus, if the VaR on an asset is $100 million at a one-week, 95% confidence level, there is a single 5% chance that the value of the asset will fall by more than $100 million over a given week. In its adapted form, the measure is sometimes defined more narrowly than the possible loss in value of "normal market risk" as opposed to all risks, which requires us to draw distinctions between normal and abnormal risk, as well as between market and non-market risk.

To calculate VaR, we need to model the portfolio (and therefore make assumptions). In particular, this involves assigning a probability to the various possible evolutions of the portfolio. VaR is therefore always conditional on modeling a hypothetical future, which necessarily has its limits.

This VaR can be used :

- To calculate economic capital.
- For monitoring market risks, both as risk reporting and as a decision-making tool.
- For credit risk.
- For regulatory requirements (regulatory reporting and specific requirements).

Key features

The VaR of a portfolio depends essentially on three parameters:

- The distribution of portfolio results. This distribution is often assumed to be normal, but many financial players use historical distributions. The difficulty lies in the size of the historical sample: if it is too small, the probabilities of high losses are not very accurate, and if it is too large, the temporal consistency of the results is lost (we are comparing results that are not comparable).
- The chosen confidence level (usually 95 or 99%). The probability that the portfolio's potential losses will not exceed the Value at Risk.
- The chosen time horizon. This parameter is very important, as the longer the time horizon, the greater the losses.

Mathematical interpretation

According to Esch, Kieffer and Lopez (in 1997) and Jorion (in 2000), the VaR of the asset under consideration, for a duration t and a probability level q is defined as the amount of loss expected so that this amount, during the period [0, t], should not be greater than the VaR with a probability of (1-q), in other words :

$$\Pr[\ Pt > VaRq\ ] = 1 - q \quad \text{equivalent to} \quad \Pr[\ Pt < VaRq\ ] = q\ .$$

Where Pt is the loss on the security at time t.

By centralizing and reducing the expression, we obtain :

$$\Pr\left\{\frac{Pt - E(Pt)}{\sigma(Pt)}\right\} \leq \frac{VaR\,q - E(Pt)}{\sigma(Pt)}$$

We can define :

$$Zq = \frac{VaRq - E(Pt)}{\sigma(Pt)}$$

VaR is therefore given by the following formula :

$$VaRq = E\,(Pt) + zq \times \sigma\,(Pt)\,.$$

- It is a synthetic indicator that gives an assessment of a portfolio's risk.
regardless of the assets that make it up.
- it's an indicator that's easy to read and interpret, even by non-specialists, even though the calculation method is highly complex.
- VaR is also a non-convex function, so merging two portfolios does not necessarily reduce risk.
- VaR indicates the maximum potential loss over a time horizon for a given confidence level. Thus, VaR gives no indication of the values taken once the threshold has been passed.

Historical VaR

The historical method only requires knowledge of the position's value in the past (e.g. historical returns for an index). In the case of a portfolio, we need to reconstitute its past value based on the price of the various assets and the current composition of the portfolio. Once the significant risk factors for the portfolio have been identified, the historical data collected is used to deduce a loss amount.

For example:

Consider a portfolio made up of several assets. To calculate the one-day historical VaR on this portfolio, we need to record all daily gains and losses over the last 1000 days (for example). Once all these data have been obtained, they need to be ranked in ascending order. To obtain the 99% VaR, simply find the 10ème (1000*(100% - 99%)) value obtained.

Benefits and challenges :

The great advantage of this method is its ease of use.

- This method requires very little calculation and technical effort.

- It requires simple techniques and little calculation.

- No need for distribution form.

- The method itself is not adaptable or derivative.

- The history must be sufficiently large compared to the VaR horizon and its confidence level, but not too large to ensure that the probability distribution has not changed too much over the period.

3.3.2.   Expected Shortfall

The subprime mortgage crisis in the United States from July 2007 onwards prompted the regulator to call for a new measure to capture risks that had not been captured. The aim is to replace VaR with Expected Shortfall (ES). The aim of this development would be to position operational risk monitoring no longer on a certain quantile of the loss distribution (per

the definition of VaR) but on the expectation of losses (P) above VaR.

If x $\in Lp$ ($\mathscr{F}$) a space Lp) is the gain of a portfolio at a future time and $0<\alpha<1$, then Expected Shortfall is defined by the following formula:

ES (Expected Shortfall) = conditional expectation of loss knowing loss > VaR

= (average size of losses in excess of VaR)

= E[loss | loss > VaR]

We can say that :

Expected Shortfall is a function of the portfolio's loss distribution, a chosen confidence level and the time horizon.

Benefits and challenges

- Expected Shortfall ESp increases as p increases.

- Encourages diversification.

- More difficult to test ex post than VaR.

3.3.3   Volatility and variability Skewness

Volatility Skewness is a measurement similar to the omega, but
using the second partial moment. It is the ratio of the upside variance to the downside variance.

$$\text{Volatility Skewness} = \frac{\sigma_U^2}{\sigma_{D2}}$$

Variability Skewness is the ratio of upside risk to downside risk.

$$\textbf{Variability Skewness} = \frac{\sigma_U}{\sigma_D}$$

3.3.4 Omega Ratio

The omega ratio is a relative measure of the probability of achieving a given return, such as a minimum acceptable return (MAR) or a target return. The greater the omega value, the greater the probability that a given return will be achieved or exceeded. Omega represents a ratio of the cumulative probability of an investment outcome above an investor's defined return level (a threshold level), to the cumulative probability of an investment outcome below an investor's threshold level. The omega concept divides expected returns into two parts - gains and losses, or returns above the expected rate (the upside) and those below (the downside). So, in simple terms, omega is seen as the ratio of rising returns to falling returns.

$$\Omega(r) = \frac{\int_r^{+\infty} (1-F(x))dx}{\int_{-\infty}^{r} F(x)}$$

- r is the target return threshold defining what is considered a gain versus a loss.
- F is the cumulative yield distribution function.

It was designed by Keating & Shadwick in 2002.

Benefits and challenges :

Since omega considers all the information available from historical investment return data, it can be used to rank potential investments in a way specific to the investor's threshold level.

However, omega decisions are not static for at least two reasons:

- As the yield information is updated, the probability distribution will change and the omega must be updated.

- As an investor's threshold level changes, the ranking between comparative investments may change.

Consequently, the omega allows investors to visualize the trade-off between risk and return at different threshold levels for different investment choices. Note that when the threshold is set to the mean of the distribution, the omega ratio is equal to 1.

3.3.5 Upside potential ratio

The Upside-potential ratio is a measure of an investment asset's performance relative to the minimum acceptable return. The measure enables a company or individual to choose investments that have performed relatively well on the downside, per unit of downside risk.

$$U = \frac{\sum_{min}^{+\infty}(R_r - R_{min})P_r}{\sqrt{\sum_{-\infty}^{min}(R_r - R_{min})P_r}} = \frac{E[(R_t - R_{min})_+]}{\sqrt{(R_t - R_{min})_-^2}}$$

Where Rr yields have been put in ascending order, Pr is the probability of yield Rr and $Rmin$ is the minimum acceptable yield.
The measurement was developed by Frank A. Sortino.

<u>**Remarks :**</u>

- The Upside-potential ratio is a measure of risk-adjusted return. All these measures depend on certain risk measures.

- In practice, the standard deviation is often used, perhaps because it is mathematically easy to manipulate. However, the standard deviation deals with deviations at-
above-average deviations (which are desirable, from the investor's point of view) in exactly the same way as he treats below-average deviations (which are less desirable, at the very least). In practice, rational investors have a preference for good returns (i.e. above-average deviations) and an aversion to poor returns (i.e. below-average deviations).

Chapter IV
Analysis and use of data

4.1 Introduction

We'll focus our study on the stocks that make up the sector-based SPDR ETFs (Exchange Traded Funds) managed by State Street Global Advisors, which track the Standard & Poor's 500 (S & P 500) index by implementing a model to explain the cross-asset returns of the equity universe using the financial ratios retained.

Cross-Sectional Analysis:

Cross-Sectional Analysis is a type of analysis that an investor, analyst or portfolio manager can perform on a company in relation to its peers within its industry or all other firms in the study universe, with the aim of assessing its performance and investment opportunities.

In a cross-sectional analysis, the analyst uses comparative measures to identify the value, debt burden, future prospects and/or operating efficiency of a target company. This enables the analyst to assess the target company's efficiency in these areas, and to make the best investment choice within a group of competitors belonging to its industry.

We will focus on the following two variables:

- PB: Price-to-Book value: the inverse of the book-to-market equity variable.
- FCPS: Free cash flow per share: this variable corresponds roughly to $C(+)/P$, except that negative cash flows are also included. The methodology applied in this work will be inspired by the literature reviewed.

Databases used

In this project, we will focus our study on the stocks that make up the SPDR sector ETFs (Exchange Traded Funds) from 15/04/2003 to 06/05/2016. The historical data required for the analyses, i.e. the financial ratios Free Cash Flow Per Share (FCFPS) and Price-to-Book value (PB) of the various companies making up the exchange traded funds in this study, were provided by the host institute "Typhoon Partners". Price data were retrieved from Yahoo Finance to calculate quarterly and annual arithmetic and logarithmic returns

31

for each stock.

Study variable dimensions

- There are two types of financial statements in the database: As Reported View (AR) and Most-Recent Reported View (MR).

AS Reported View (AR)

- Excludes restatements.
- Point of view of time with data time indexed to the date the Form 10 regulatory filing was submitted to the SEC.
- Presents data for the last reporting period at this filing date.
- May include several observations in a quarter if a deposit is also made during the quarter.
- On limited occasion, may not have observations in a given quarter. Sometimes, companies are delayed in reporting by up to 18 months.
On these occasions, they may report several documents to be caught up to the same date, in which case these data sets will provide the most recent reporting period.
- Generally suitable for backtesting.

Most-Recent Reported view (MR):

- Includes restatements.
- Time indexed to the financial reporting period.
- Presents the most recently reported data for this reporting period.
- Generally appropriate for assessing company performance after adjustments for mergers and disposals.

In addition, we will focus on the following two temporal dimensions:

- Annual (Y): annual observations.
- Quarterly (Q): quarterly observations (available only for US companies, not for foreign companies).

Dimensions	As Reported View	Most-Recent Reported View
Annual	ARY	MRY
Quarterly	ARQ	MRQ

The sectors that make up Exchange Traded Funds :

SPDR Exchange Traded Funds are made up of the following 11 sectors:

1. Consumer Discretionary sector rated XLY.

2. Consumer Staples sector rated XLP

3. Energy sector (Energy) rated XLE.

4. Financial Services sector rated (XLFS).

5. Financial sector (Financial) rated XLF.

6. Health Care sector (Health Care) rated XLV.

7. Industrial product sector (Industrial) rated XLI.

8. Materials sector rated XLB.

9. Real Estate sector rated XLRE.

10. Technology sector (technology)rated XLK.

11. Utilities sector, rated XLU.

4.2 Analysis of quarterly and annual returns for each sector

Table 1-a shows the average quarterly and annual arithmetic returns, as well as the number of firms in each sector.

Table1-a. Industries sorted by average quarterly and annual arithmetic returns from 15/04/2003 to 06/05/2016 (48665 observations).

Industries	Average quarterly arithmetic returns	Average annual arithmetic yields	Number of firms
XLE	0.021	0.078	37
XLB	0.024	0.110	27
XLF	0.033	0.105	92

XLFS	0.050	0.127	64
XLI	0.025	0.094	68
XLK	0.033	0.167	73
XLP	0.024	0.100	36
XLRE	0.026	0.105	28
XLU	0.021	0.077	28
XLV	0.031	0.132	57
XLY	0.032	0.145	87

Quarterly returns :

The best-performing sector in terms of quarterly returns is the financial services sector (XLFS), with a return of 0.050 (Fig. 1a). In second place are the technology (XLK) and financial (XLF) sectors, with quarterly arithmetic returns of 0.033. Note that the Energy sector (XLE) and the Utilities sector (XLU) are in last place, with an average quarterly return of 0.021 (Fig. 1a).

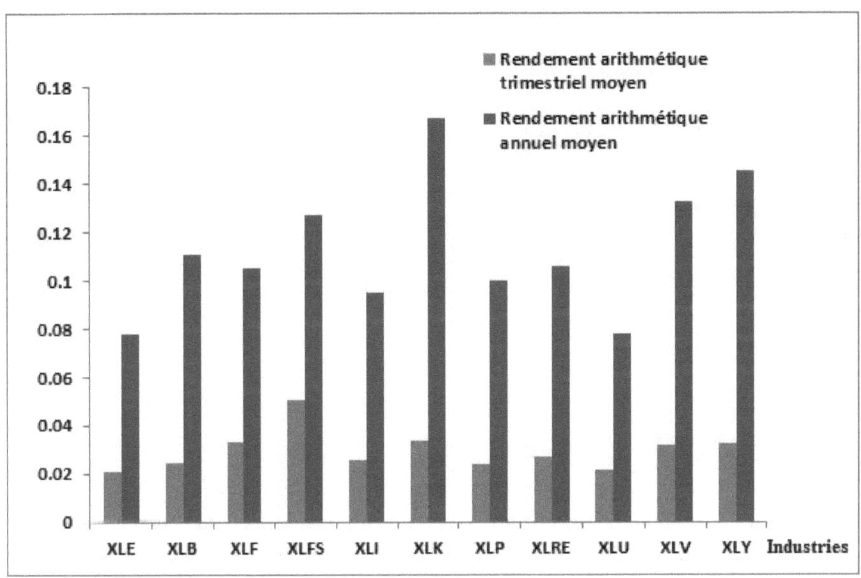

Fig. 1a. Average quarterly and annual arithmetic industry yields from 15/04/2003 to 06/05/2016 (48665 observations).

Annual yields :

In terms of annual returns, we find that the Technology sector is in first place with an annual arithmetic return of 0.167 (Fig. 1a), while the Consumer Discretionary sector is in second place with an annual arithmetic return of 0.145. As with the quarterly returns, we find that the last place goes to the Energy and Utilities sectors, with a return of around 0.077.

4.3 Analysis of financial ratios

Table 1-b and Figure 1b show the various financial ratios FCFPS "Quarterly (MRQ) and annual (MRY) Free Cash Flow Per Share and PB Price-to-Book Value, sorted by sector.

Table 1-b. Industries sorted by FCFPS and average PB ratios in Most-Recent Reported View dimension from 15/04/2003 to 06/05/2016 (48665 observations).

Industries	FCFPS_MRQ	FCFPS_MRY	PB_MRQ	PB_MRY
XLE	0.108	0.473	2.463	2.378
XLB	0.638	2.616	3.689	3.597
XLF	1.170	4.716	2.060	2.773
XLFS	2.201	8.937	1.253	2.952
XLI	0.845	3.361	0.523	0.299
XLK	0.704	2.751	4.668	5.493
XLP	0.646	2.579	7.222	7.382
XLRE	0.091	0.258	3.057	3.032
XLU	0.850	3.360	0.520	0.300
XLV	0.967	3.780	4.282	4.347
XLY	0.725	2.895	8.083	4.656

Free Cash Flow Per Share (FCFPS):

The financial services sector (XLFS) has the highest values of average quarterly and annual ratios, FCFPS_MRQ and FCFPS_MRY, which are equal to 2,201 and 8,937, respectively. In second place, we find the finance sector (XLF) with quarterly and annual ratios of 1,170 and 4,716, respectively. In last place is the real estate sector (XLRE), with quarterly and annual ratios of 0.091 and 0.257 respectively.

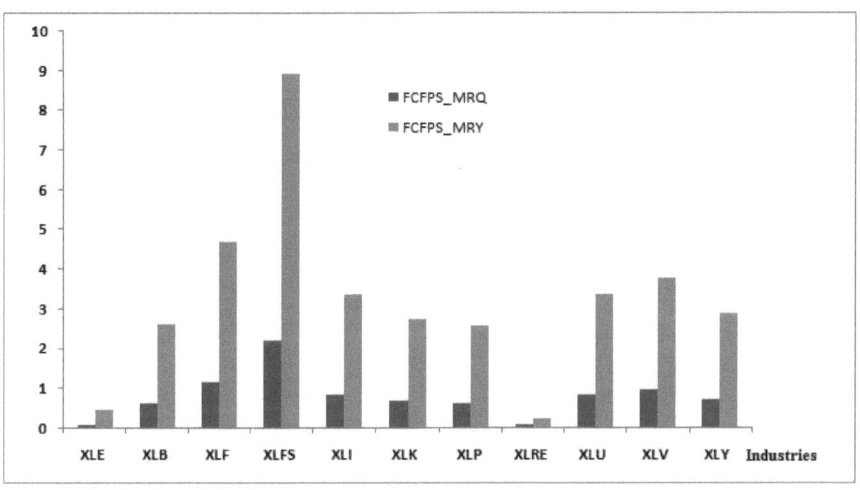

Fig.1b. Values of quarterly (MRQ) and annual (MRY) FCFPS financial ratios, in Most Recent Reported View dimension, sorted by sector from 15/04/2003 to 06/05/2016 (48665 observations).

Price-to-Book value (PB):

The Consumer Discretionary sector has the highest average quarterly Price-to-Book ratio, equal to 8.083 (Fig. 1c), followed by the Consumer Products sector (XLP) with a PB_MRQ equal to 7.222. The lowest PB_MRQ values, equal to 0.520 and 0.523, correspond to the Industrial Products (XLU) and Utilities (XLI) sectors, respectively. As for the average PB_MRY ratio, the highest value is for the Consumer Products sector (XLP), equal to 7.382, followed by the Technology sector (XLK) with a value of 5.493. The industrial products sector (XLI) occupies the last position, with a PB_MRY ratio equal to 0.299.

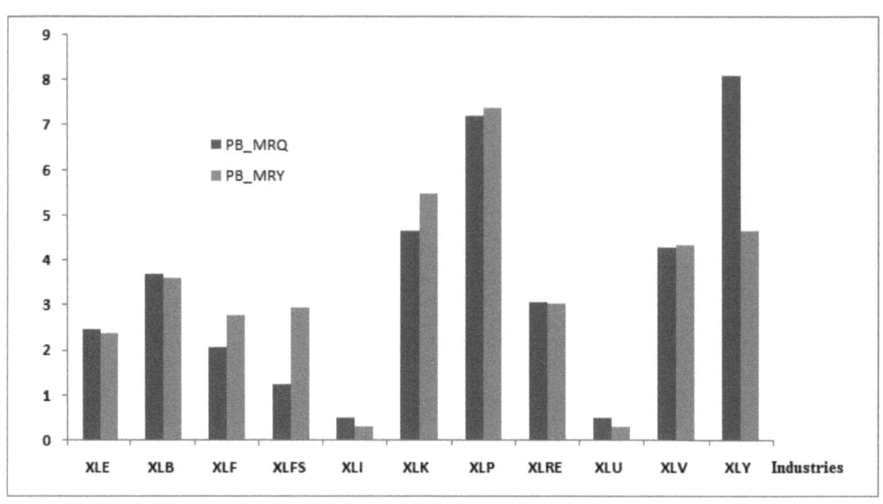

Fig.1c. Values of quarterly (MRQ) and annual (MRY) PB financial ratios, in Most Recent Reported View dimension, sorted by each sector from 15/04/2003 to 06/05/2016 (48665 observations).

Table 1-c. Industries sorted by FCFPS and average PB ratios in dimension As Reported View from 15/04/2003 to 06/05/2016 (48665 observations).

Industry	FCFPS_ARQ	FCFPS_ARY	PB_ARY	PB_ARQ
XLE	0.082	0.422	2.424	2.500
XLB	0.7045	2.611	3.666	3.620
XLF	1.128	4.566	2.716	2.170
XLFS	1.623	6.661	2.164	1.429
XLI	0.887	3.382	3.467	1.777
XLK	0.713	2.726	6.016	4.916
XLP	0.663	2.563	7.799	7.348
XLRE	0.102	0.258	3.080	3.150

XLU	0.597	2.318	2.901	1.735
XLV	0.983	3.783	4.411	4.419
XLY	0.774	2.857	4.999	8.594

Free Cash Flow Per Share (FCFPS) in dimension As Reported View:

The financial services sector (XLFS) has the highest average values of FCFPS_ARQ and FCFPS_ARY ratios, which are equal to 1.623 and 6.661, respectively (Fig. 1d). In second place, we find the finance sector (XLF) with ratios of 1.128 and 4.566. Last place is reserved for the energy sector (XLE) with a value of 0.082 for quarterly average ratios and the real estate sector (XLRE) with a value of 0.258 for annual average ratios (Fig. 1d).

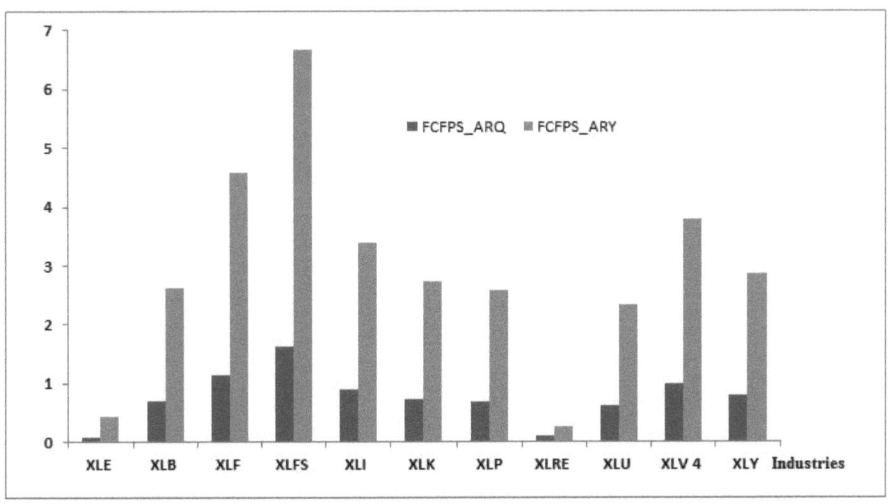

Fig.1d. Values of quarterly (ARQ) and annual (ARY) FCFPS financial ratios in As Reported View dimension, sorted by each sector from 15/04/2003 to 06/05/2016 (48665 observations).

Price-to-Book value (PB) in dimension As Reported View:

The Consumer Discretionary sector (XLY) has the highest average quarterly Price-to-Book Ratio value, equal to 8.594 (Fig. 1e), followed by the Consumer Products sector (XLP) with an average PB_ARQ ratio equal to 7.348. The lowest PB_ARQ ratio value, equal to 1.429, relates to the Financial Services sector (XLFS). The lowest PB_ARQ ratio value, equal to 1.429, relates to the financial services sector (XLFS).

As for the average annual PB_ARY ratio, the highest value corresponds to the consumer products sector (XLP), with a value of 7.799. In second place, we find the technology sector (XLK) with a value of 6.016. The last position is occupied by the industrial products sector (XLFS) with a PB_ARY ratio equal to 2.164.

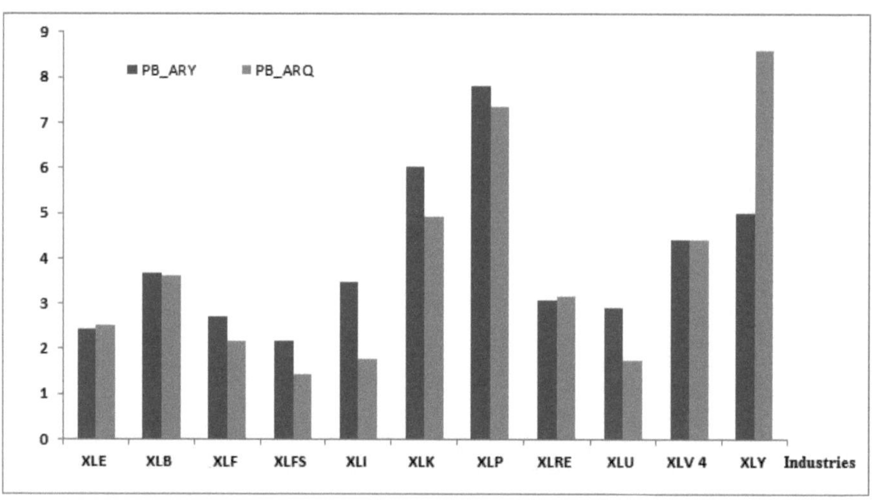

Fig.1e. Values of quarterly (ARQ) and annual (ARY) PB financial ratios in As Reported dimension sorted by each sector from 15/04/2003 to 06/05/2016 (48665 observations).

4.4 Various risk measures

In this section, we present the various analyses and values relating to the risk measures studied during the course.

4.4.1   Value at Risk (VaR)

Table 2.a shows the different" Value at Risk" quarterly and annual realized returns sorted by sector. For the calculation method, we have focused on historical VaR (with p=0.95).

Table 2.a Industries sorted by Value at Risk (p=0.95)

Industries	XLB	XLY	XLF	XLV	XLK	XLRE	XLU	XLI	XLE	XLP	XLFS
Yields quarterly	-0.29	-0.26	-0.27	-0.21	-0.24	-0.23	-0.21	-0.23	-0.32	-0.20	-0.27
Yields annual	-0.50	-0.51	-0.51	-0.38	-0.50	-0.39	-0.38	-0.37	-0.55	-0.36	-0.48

The Consumer Products sector (XLP) has the most interesting" Value at Risk" for quarterly returns, which is equal to -0.20, meaning that with a 95% probability the quarterly return for the Consumer Products sector cannot be less than -0.20. The healthcare sector (XLV) occupies second position, with a VaR equal to -0.21, while the utilities sector (XLU) occupies third position, with a VaR of -0.21. The last position is devoted to the Utilities sector (XLE), with a VaR of
-0.32. Looking at annual arithmetic returns, the Consumer Products sector (XLP) has the most interesting VaR, equal to -0.36. Next, we find the healthcare sector (XLI), followed by both the utilities (XLU) and industrial products (XLV) sectors, and after the real estate sector (XLRE) with a Value at Risk value equal to -0.37, -0.38 and -0.39, respectively. While the last position is occupied by the energy sector (XLE) with a VaR equal to -0.55 (Fig. 2a).

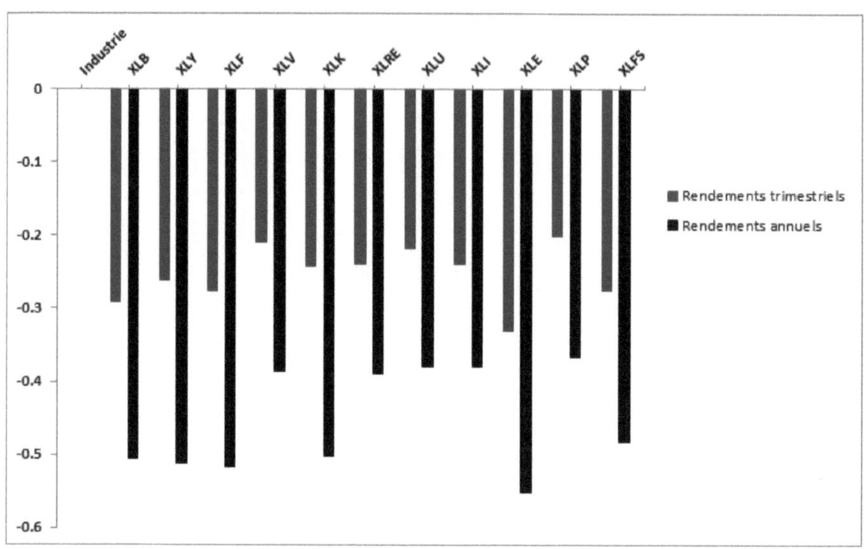

Fig. 2a. Quarterly and annual returns for industries sorted by Value at Risk (p=0.95).

4.4.2 Expected Shortfall (ES)

Table 2.b shows the different ES of quarterly and annual realized returns sorted by sector. For the calculation method, we focused on the historical ES (with p=0.95).

Table 2.b Industries sorted by Expected Shortfall (p=0.95)

Industries	XLB	XLY	XLF	XLV	XLK	XLRE	XLU	XLI	XLE	XLP	XLFS
Yields quarterly	-0.42	-0.38	-0.42	-0.34	-0.36	-0.41	-0.32	-0.34	-0.42	-0.31	-0.40
Yields annual	-0.66	-0.62	-0.66	-0.50	-0.63	-0.62	-0.50	-0.51	-0.63	-0.50	-0.65

The Consumer Products (XLP) sector has the most interesting value of all.

"Expected Shortfall" of quarterly returns equal to -0.31. In second and third place, we find the Utilities sector (XLU) and the Healthcare sector (XLV) with ES values equal to -0.32 and -0.34, respectively. The lowest ES values are found in the Financial sector (XLF), which is equal to -0.42. With regard to annual returns, we obtained an ES value equal to -

0.50 for the best value achieved by the consumer products sector (XLP), the "Utilities" sector XLU and the healthcare sector XLV, while the last position is occupied by the financial sector (XLB) with a value of -0.66 (Fig. 2b).

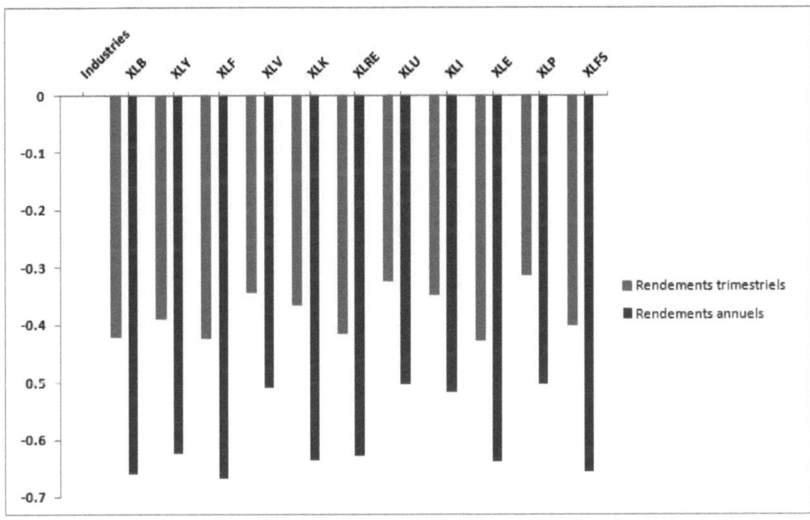

Fig. 2b. Quarterly and annual returns for industries sorted by Expected Shortfall.

4.4.3   Omega

Table 2.c shows the various Omega values (quarterly and annual returns) sorted by sector. We can think of omega as the ratio of increasing returns to decreasing returns. We find that the energy sector (XLE) has the best omega value, which is the smallest, equal to 3.44 and 3.55 for quarterly and annual returns respectively, while the healthcare sector (XLV) admits the worst omega values (4.19 and 5.38) respectively.

Table 2.c Industries sorted by Omega

Industries	XLB	XLY	XLF	XLV	XLK	XLRE	XLU	XLI	XLE	XLP	XLFS
Yields quarterly	3.82	3.74	3.50	4.19	3.75	3.76	3.83	3.98	3.44	3.90	3.48
Yields annual	4.17	4.22	3.87	5.38	4.31	4.35	4.50	4.59	3.55	5.40	3.83

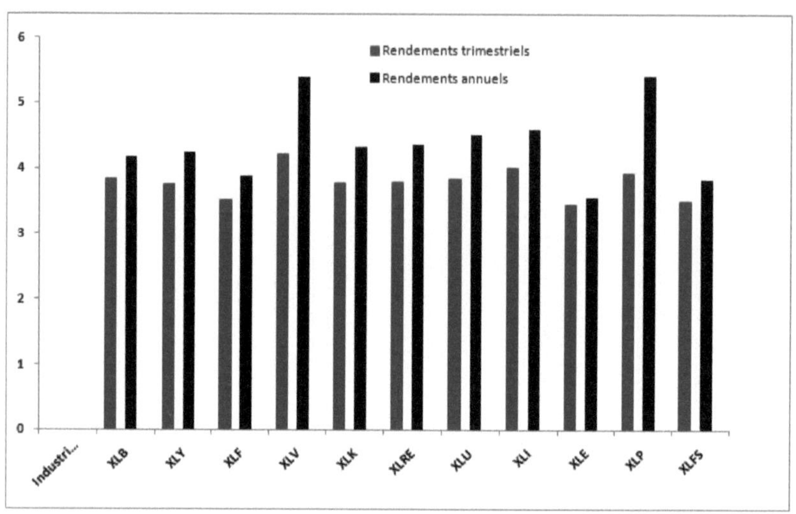

Fig. 2c Quarterly and annual yields for industries sorted by Omega.

4.4.4   Volatility Skewness

Figure 2d and Table 2.d show the different Skewness volatility values for quarterly and annual arithmetic returns sorted by sector. We note that the energy sector (XLE) has the lowest Skewness Volatility value, equal to 1.13 for quarterly returns, and the highest Skewness Volatility value for annual returns is reserved for the "Utilities" sector (XLU), with a value of 2.60. The financial sector (XLF), on the other hand, has the lowest Skewness Volatility value, equal to 1.13 for quarterly returns. Whereas the Financials (XLF) and Financial Services (XLFS) sectors have very high Skewness volatility values (31.1 and 43.12 for quarterly returns, and 15.5 and 17.8 for annual returns respectively). This is due to the various financial crises in this period, such as the global financial crisis that began in 2007, which amplified the movement and caused stock market prices (especially for banks and financial institutions) to plummet.

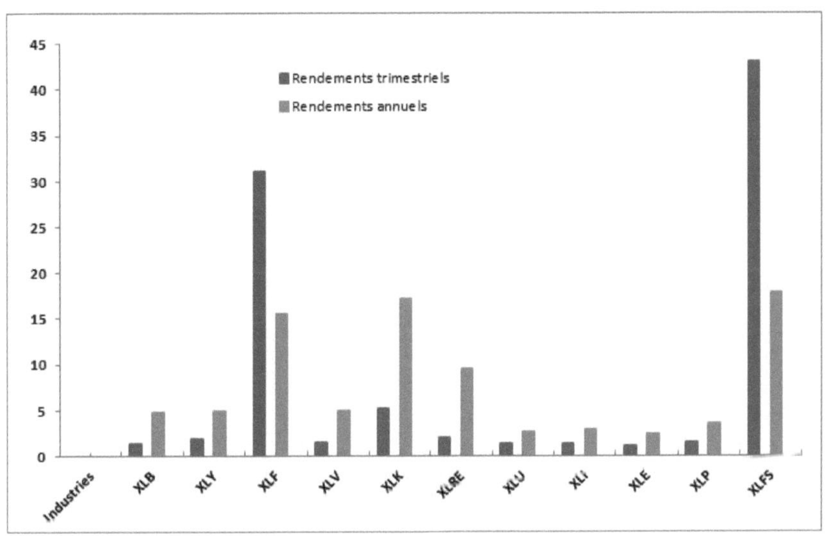# Industries sorted by Skewness volatility

Fig. 2d. Quarterly and annual yields for industries sorted by Skewness volatility.

Table 2.d Industries sorted by Skewness volatility.

Industries	XLB	XLY	XLF	XLV	XLK	XLRE	XLU	XLI	XLE	XLP	XLFS
Yields quarterly	1.31	1.87	31.10	1.42	5.19	1.94	1.34	1.28	1.13	1.41	43.12
Yields annual	4.78	4.86	15.56	4.85	17.14	9.51	2.60	2.83	2.31	3.52	17.80

4.4.5   Upside potential ratio

The Upside Potential Ratio is a measure of the return on an investment asset in relation to the Minimum Acceptable Return (MAR). Table 2.e shows the different Upside potential ratios for quarterly and annual realized returns, sorted by sector. Concerning quarterly returns, the financial services sector (XLFS) has the highest value, equal to 1.01 (Fig. 2e), followed by the technology (XLK) and financial sectors with values of 0.82 and 0.82, and lastly the real estate sector (XLRE) with a value of 0.69. For annual returns, the technology sector (XLK) has the highest ratio, equal to 1.30, followed by the financial services sector (XLFS) with a value of 1.22, while the consumer products sector (XLP) is in last place with a value of 0.88.

Table 2.e Industries sorted by Upside potential ratio

Industries	XLB	XLY	XLF	XLV	XLK	XLRE	XLU	XLI	XLE	XLP	XLFS
Yields quarterly	0.70	0.81	0.82	0.72	0.82	0.69	0.72	0.71	0.76	0.71	1.01
Yields annual	1.04	1.18	1.06	1.06	1.30	0.97	0.91	0.98	1.04	0.88	1.22

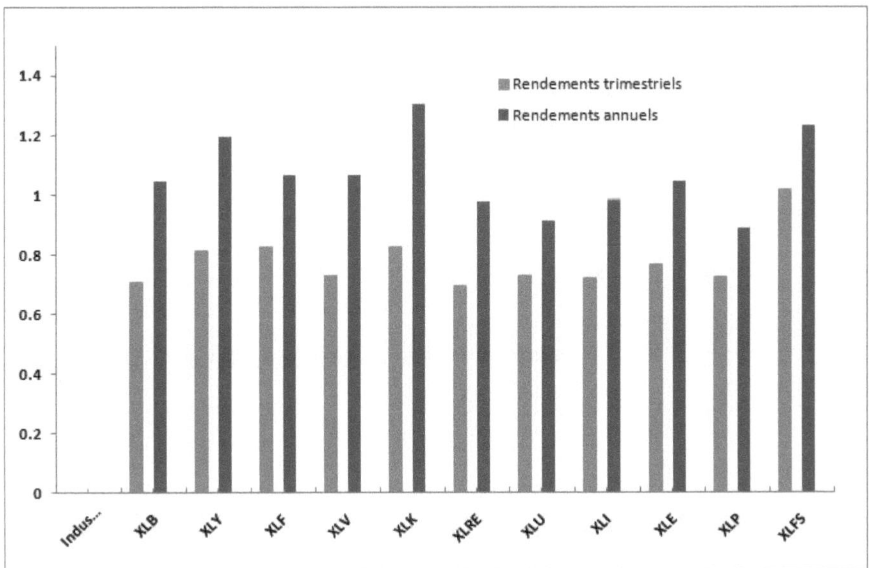

Fig. 2d Quarterly and annual industry yields sorted by Upside potential ratio

4.5 Analysis of data

The aim of the analysis is to test whether a better approximation of information o n future returns is obtained by breaking the study variable, in our case (PB and FCFBS), into two components:

- The first component is the difference between the measure of the variable X for firm i at time t and the average value for the sector (this difference is called the within-industry measure).
- The second component is the across-industry measure of the variable.

47

To formally examine the importance of within-industry measure and across-industry measure variables in predicting expected stock returns, we start with the Fama-MacBeth (FM) regression, which is a simple panel regression.

Specification:

Rit = yAt + yBt Xit +εit (1)

- Rit is the yield of firm i at time t.
- Xit refers to the measurement of variable X for company i at time t.
- yAt and yBt are regression coefficients.

We modify this specification as follows: f

Rit = γ0t + γ1t XIit + γ2t (Xit - XIit)+εit   (2)

- XIit is the equiponderated average characteristic of companies in the company's industry.

For example, X could be the company's price-to-book value (PB).

In Equation (1), the size of the average time series of y^Bt, □B̄t, provides the classic FM test of whether expected returns are unconditionally related to the book-to-market ratio.

In equation (2), the significance of □□Ῑt, tests whether distressed companies have higher expected returns than companies in growth sectors, the significance of □□2̄t, tests whether companies that look distressed relative to their industry have
We refer to Xit as the market measure of variable X, the average value of variable X for all companies within the industry denoted XIit, and the deviation from XIit. We refer to Xit as the market measure of variable X, the average value of variable X for all companies within the industry noted XIit, (across-industry measure of variable X) and the deviation from the industry average value (Xit - XIit) called (within-industry measure). Generally speaking, our FM tests show that within-industry measurement of variables provides coefficients of slightly greater magnitude than market-wide regressions, and with smaller

standard errors.

Fama-MacBeth Regressions:

In this section, we assess the power of variables across its industry and across several industries to explain cross-asset and equity universe returns. To compare our industry measures to the standard specification of equation **(1)**. Table 3 shows the mean slopes and t-values of equations **(1)**, **(2)** and
(3) for each variable.

4.5.1 Most Recent Reported View **PB_MRQ** dimensional data analysis and

arithmetic performance :

In this paragraph, we are interested in quarterly arithmetic equity returns (as the variable to be explained) and the study variable is Price-to-Book value, the observations are quarterly of quarterly duration in Most Recent Reported View dimension.

Like the market value of equity, Price-to-Book value is a reliable price (Stattman [1980], Rosenberg, Reid and Lanstein [1985], Bondt and Thaler [1987], Fama and Frensh [1992]), but the within-industry version offers a better measure of the relationship between Price-to-Book value and quarterly stock market returns.

Table 3: Presentation of Market-wide, Across-industry and Within-industry coefficients with arithmetic yield as the variable to be explained.

Market-Wide regression equation:

(1) $R_{it} = y_{At} + y_{Bt} X_{it} + \varepsilon_{it}$.

Industry regression equation :

(2) $R_{it} = y_{0t} + y_{1t} X_{Iit} + y_{2t} (X_{it} - X_{Iit})$.

Alternate Industry regression equation :

(3) $R_{it} = y_{0t} + (y_{1t} - y_{2t}) X_{Iit} + y_{2t} X_{it}$

- Xlit refers to the average value for the sector.
- Rit is the yield of firm i at time t.
- Xit refers to the measurement of variable X for company i at time t.

Arithmetic yield	Market-wide	Across -industry	Within- industry
FCFPS_MRQ (1)	5.917 e-03 (65.31)		
(2)		7.565e-03 (8.43)	5.900e-03 (64.78)
(3)	5.900e-03 (64.78)	1.665e-03 (1.84)	
FCFPS_MRY (1)	8.453e-03 (159.90)		
(2)		-1.113e-03 (-3.38)	8.707e-03 (162.57)
(3)	8.707e-03 (162.57)	-9.819e-03 (-29.48)	
PB_MRQ (1)	1.858e-05 (4.96)		
(2)		5.388e-04 (3.66)	1.824e-05 (4.87)
(3)	1.824e-05 (4.87)	5.205e-04 (3.54)	
PB_MRY (1)	7.487e-05 (7.41)		
(2)		1.054e-02 (34.60)	6.338e-05 (6.28)
(3)	6.338e-05 (6.28)	1.048e-02 (34.37)	

The average within-industry slope is slightly lower than the average market-wide slope (1.824e-05 vs. 1.858e-05), while the within-industry t-value (4.87) is also slightly lower than the market-wide t-value (4.96). This is due to the large Across-industry coefficient value which is equal to 5.388e-04 with t-value equal to 3.66.

Estimates of equation (3) show that the difference between the within-industry coefficient and the across-industry coefficient is statistically significant. The difference of 5.205e-04 has a t-value of 3.54.

PB_MRY and arithmetic yield :

In this section, we analyze annual arithmetic stock returns (as the variable to be explained) and the study variable is Price-to-Book value, with annual one-year observations in the

Most Recent Reported View dimension.

The mean value of within-industry slope (6.338e-05) is slightly smaller than the mean value of Market-wide slope (7.487e-05), with the t-value of within-industry slope (6.28) smaller than the t-value of Market-wide slope (7.41). This decrease in power is driven by a strong effect of the across-industry slope 1.054e-02 with t-value of 34.60. As in the case of individual stocks, winning industries over the past year have higher expected returns than losing industries.

FCFPS_MRQ and arithmetic efficiency :

As in the previous section, in this section we focus on the relationship between quarterly arithmetic stock returns (as the variable to be explained) and the study variable is Free cash flow per share, observations are quarterly of quarterly duration in Most Recent Reported View dimension.
The average within-industry slope is equal to the average market-wide slope (5.900e-03), with the t-value of the market-wide slope slightly greater than the t-value of the within-industry slope (65.31 and 64.78 respectively). The results of specification (3) show a difference between the within-industry coefficient and the across-industry coefficient of 1.665e-03 with a t-value of (1.84), which is not statistically significant since it is less than 1.96.

FCFPS_MRY and arithmetic efficiency :

In this case, we work with annual arithmetic stock returns (as the variable to be explained) and the study variable is Free cash flow per share, with annual observations lasting one year from Most Recent Reported View. We find, in line with previous results [1], that the average within-industry slope, equal to 8.707e-03, is slightly higher than its average market-wide slope, equal to 8.453e-03, with the t-value of Within-industry slope and greater of market-wide slope (162.57 vs 159.90). Furthermore, the estimates of equation (3) show that the difference between the Within-industry coefficient is statistically significant. The difference is 1.665e-03 with a t-value of 1.84.

PB_MRQ and logarithmic efficiency :

We are interested in quarterly logarithmic equity returns (as the variable to be explained) and the study variable is Price-to-Book value, observations are quarterly of quarterly duration in Most Recent Reported View dimension.

The average Within-industry slope, equal to 1.892e-05, is slightly lower than the average Market-wide slope, equal to 2.057e-05, with the Market-wide t-value greater than the Within-industry slope t-value (11.29 vs. 10.38). This decrease in power is driven by a strong effect of the across-industry slope measure 2.622e-03 with a t-value of 36.26 (Table 4).

Estimates of equation (3) show that the difference between the Within-industry coefficient is statistically significant. This difference is 1.593e-02 with a t-value of 70.20.

Table 4: Presentation of Market-wide, Across-industry and Within-industry coefficients when logarithmic yield is the variable to be explained.

Housing yield		Market-wide	Across -industry	Within- industry
FCFPS_MRQ	(1)	1.277e-03 (30.57)		
	(2)		-4.114e-03 (-11.80)	1.355e-03 (32.22)
	(3)	1.355e-03 (32.22)	-5.470e-03 (-15.58)	
FCFPS_MRY	(1)	1.667e-03 (48.61)		
	(2)		-1.583 e-03 (-8.63)	1.784 e-03 (51.13)
	(3)	1.784 e-03 (51.13)	-3.368 e-03 (-18.05)	
PB_MRQ	(1)	2.057e-05 (11.29)		
	(2)		2.622e-03 (36.26)	1.892e-05 (10.38)
	(3)	1.892e-05 (10.38)	2.603e-03 (35.99)	
PB_MRY	(1)	7.285e-05 (9.67)		
	(2)		1.596e-02 (70.58)	5.135e-05 (6.85)
	(3)	5.537e-05 (7.36)	1.593e-02 (70.20)	

PB_MRY and logarithmic yield:

In this paragraph, we work with quarterly logarithmic stock returns (as the variable to be explained) and the study variable is Price-to-Book value, the observations are annual with a duration of one year in dimension Most Recent Reported View). Thanks to the strong effect of the measure across-industry slope 1.596e-02 with t-value of 70.58 we find that the average Within-industry slope, equal to 5.135e- 05, is lower than the average Market-wide slope, equal to 7.285e-05 with t-value also lower (6.85 vs. 9.67). Estimates of equation (3) show a statistically significant difference between within-industry and across-industry coefficients of 1.593e-02 with a t-value of (70.20).

FCFPS_MRQ and logarithmic efficiency:

We note that the average within-industry slope, equal to 1.355e-03, is slightly higher than the average market-wide slope, equal to 1.277e-03, with the Within-industry Market-wide t-value greater than the Market-wide slope t-value (32.22 vs. 30.57). We also note that the difference between the within-industry coefficient and the across-industry coefficient is statistically significant -5.470e-03 with t-value equal to -15.58.

FCFPS_MRY and logarithmic yield:

In line with the results found in the previous paragraph, we find that the average within-industry slope is higher than the average market-wide slope (1.784 e-03 vs. 1.667e-03), with t-value also higher (51.13 vs. 48.61). We also find that the difference between the within-industry coefficient and the across-industry coefficient is statistically significant -3.368 e-03 with t-value equal to -18.05.

4.5.2   As Reported data analysis

In this section, we continue with the same work as before, but with the study variables are As Reported.

FCFPS_ARQ and arithmetic efficiency:

Table 5 shows that the average within-industry slope equals the average market-wide slope (3.100 e-03), with the t-value of the market-wide slightly greater than the t-value of the within-industry slope (44.60 vs. 44.14). The results of specification (3) show a statistically significant difference between the within-industry and across-industry coefficients of 3.847e-03 with a t-value of (4.19).

Table 5: Presentation of Market-wide, Across-industry and Within-industry coefficients with arithmetic yield as the variable to be explained.

Arithmetic yield		Market-wide	Across -industry	Within- industry
FCFPS_ARQ	(1)	3.100 e-03 (44.60)		
	(2)		6.955e-03 (7.61)	3.100 e-03 (44.14)
	(3)	3.100 e-03 (44.14)	3.847e-03 (4.19)	
FCFPS_ARY	(1)	6.773e-03 (128.30)		
	(2)		-1.299e-03 (-3.42)	6.989e-03 (129.13)
	(3)	6.989e-03 (129.13)	-8.288e-03 (-21.66)	
PB_ARQ	(1)	1.928e-05 (5.44)		
	(2)		4.398e-04 (2.99)	1.904e-05 (5.37)
	(3)	1.904e-05 (5.37)	4.208e-04 (2.86)	
PB_ARY	(1)	3.799e-05 (3.64)		
	(2)		1.054e-02 (34.60)	2.722e-05 (2.61)
	(3)	2.722e-05 (2.61)	1.152e-02 (33.80)	

FCFPS_ARY and arithmetic efficiency:

We note that the average Within-industry slope, equal to 6.989e-03, is slightly higher than its average Market-wide slope, equal to 6.773e-03, with a higher Within-industry slope t-value than Market-wide slope (129.13 vs. 128.30). Furthermore, the estimates of equation (3) show that the difference between the Within-industry coefficient is statistically significant. The difference is -8.288e-03 with a t-value of -21.66.

PB_ARQ and arithmetic yield:

In this paragraph, we are interested in quarterly arithmetic equity returns (as the variable to be explained) and the study variable is Price-to-Book value, the observations are quarterly of quarterly duration in the As Reported dimension.

The average within-industry slope and the average market-wide slope are equal to 1.904e-05, and so are the t-values, which are equal to 5.37.

Estimates of equation (3) show that the difference between the within-industry coefficient and the across-industry coefficient is statistically significant. The difference is 4.208e-04 with a t-value of 2.86.

PB_ARY and arithmetic yield :

The mean value of within-industry slope is slightly smaller than the mean value of market-wide slope (2.722e-05 vs. 3.799e-05), with a t-value of within-industry slope (2.61) smaller than the t-value of market-wide slope (3.64). This decrease in power is driven by a strong effect of the across-industry slope measure 1.152e-02 with a t-value equal to (33.80).

PB_ARQ and logarithmic yield:

In this paragraph, we are interested in quarterly logarithmic equity returns (as the variable to be explained) and the study variable is Price-to-Book value, the observations are quarterly of the quarterly duration of dimension As Reported.

Table 6 shows that the average Within-industry slope is slightly lower than the average Market-wide slope (2.035e- 05 vs. 2.158e-05), with the Market-wide t-value greater than the Within-industry slope t-value (13.22 vs. 12.47), driven by a strong across-industry slope effect of 2.050e-03 with a t-value of 30.85 (Table 6). Equation (3) shows that the difference between across-industry slope and Within-industry slope is equal to 2.030e-03 with a t-value of 30.53.

Table 6: Presentation of Market-wide, Across-industry and Within-industry coefficients in the As Reported dimension, where logarithmic yield is the variable to be explained.

Housing yield		Market-wide	Across -industry	Within- industry
FCFPS_ARQ	(1)	6.561e-04 (20.25)		
	(2)		-4.222e-03 (-10.96)	6.909e-04 (21.25)
	(3)	6.909e-04 (21.25)	-4.913e-03 (-12.71)	
FCFPS_ARY	(1)	1.218e-03 (36.08)		
	(2)		-2.297e-03 (-11.56)	1.333e-03 (38.91)
	(3)	1.333e-03 (38.91)	-3.969e-03 (-19.70)	
PB_ARQ	(1)	2.158e-05 (13.22)		
	(2)		2.050e-03 (30.85)	2.035e-05 (12.47)
	(3)	2.035e-05 (12.47)	2.030e-03 (30.53)	
PB_ARY	(1)	4.330e-05 (6.54)		
	(2)		1.395e-02 (67.30)	2.915e-05 (4.40)
	(3)	2.915e-05 (4.40)	1.392e-02 (67.12)	

PB_ARY and logarithmic yield:

Thanks to the strong effect of the across-industry slope measure 1.395e-02 with t-value of (67.30) we find that the average within-industry slope, equal to 2.915e-05, is lower than the average market-wide slope which is equal to 4.330e-05 with t-value is also lower (4.40 vs. 6.54). Estimates of equation (3) show a statistically significant difference between within-industry and across-industry coefficients of 1.392e-02 with a t-value of (67.12).

FCFPS_ARQ and logarithmic yield:

The average within-industry slope, equal to 6.909e-04, is slightly higher than the average market-wide slope which, equal to 6.561e-04 with the within-industry t-value, is greater than the market-wide slope t-value (21.25 vs. 20.25). We also note that the difference between the within-industry coefficient and the across-industry coefficient is statistically significant -4.913e-03 with t-value equal to -12.71.

FCFPS_ARY and logarithmic yield:

In line with the results found in the previous paragraph, we find that the average within-industry slope is higher than the average market-wide slope (1.333e-03 vs. 1.218e-03) with t-value also higher (38.91vs 36.08). We also find that the difference between the within-industry coefficient and the across-industry coefficient is statistically significant -3.969e-03 with t-value equal to -19.70.

Chapter V
Extension to investment strategies and portfolio insurance

5.1 Motivation

Sector SPDRs are subject to risks similar to those of equities, including those relating to short selling and margin accounts. All ETF's are subject to risks, including potential loss of capital. Sector ETF's are also subject to non-diversified sector risk, which will result in greater price fluctuations than the overall market.

In this chapter, we compare a number of investment strategies, based on the statistical results obtained previously, as well as portfolio insurance.

5.2. Comparison of some investment strategies
5.2.1   Equal Sector strategy

An Equal Sector strategy is one that offers exposure to the large-cap equity market by investing in equal proportions in the SPDR sector, so investors shouldn't double-count the financial sector.

A sound equi-weighted strategy can include either XLF or XLFS and XLRE in place of XLF, as well as the other eight sectors. This strategy offers moderate but significant exposure to all market sectors.

As a result, investors have the opportunity not only to participate in a sector rally, but also to minimize the negative impact of a crash in a particular sector by rebalancing to quarterly equal weight. In addition, the Equal Sector strategy offers the following advantages:

- Diversification.
- The opportunity to participate in a market gathering in all sectors.
- Transparency and control over sector allocation.
- Reducing the negative impact of an accident in a particular sector.
- Lower volatility than the S & P 500 index.

5.2.2   Strategies tax

Select Sector SPDRs can help investors manage the tax impact of investment portfolios more effectively.

<u>Strategy 1</u>

Sell individual trading stock positions currently below the purchase price, realize the loss and maintain similar sector exposure by purchasing the appropriate Select Sector SPDR.

Example:

You currently manage shares held by a Select Sector that are trading below their original purchase price.

You can sell your loss-making positions, realize losses, and buy the Select Sector SPDR that holds the stock positions to maintain exposure to these and other stocks in that sector.

<u>Strategy 2:</u> **Take losses in funds currently held while maintaining sector exposure**

Sell a currently traded ETF, closed-end fund or mutual fund below the purchase price, realize the loss and maintain similar sector exposure by buying the appropriate Select Sector SPDR.

Example:

You currently hold TechnologyFund shares, which are trading below your initial purchase price. You can sell your loss position, realize the loss, and buy the Technology Select Sector SPDR (XLK) to maintain similar exposure.

<u>Strategy 3</u>: Maintain a customized Select Sector SPDR portfolio and rebalance at year-end to harvest tax losses.

Select an SPDR sector that allows us to buy the S & P 500 index in chunks, giving you greater flexibility to customize a portfolio and better manage after-tax returns.

Example:

If our objective is income or growth, we can sell the sectors of unrealized losses at year-end and maintain exposure to these sectors by buying similar ETFs. After waiting at least 31 days, we can rebalance the portfolio by switching back into the appropriate Select Sector SPDRs with weightings appropriate to our investment objective.

5.2.3   Hedging strategies

<u>Strategy 1</u>: Using the flexibility of Select Sector SPDRs offsets long-term losses in Select Sector SPDRs that match your positions in individual stocks, and reduces downside risk without realizing a taxable event on your long positions in these stocks.

Example:

We hold significant positions in A-shares and B-shares with unrealized gains. To hedge the downside risk without recognizing a taxable event, we sell the short Select Sector SPDR which includes these companies.

<u>Strategy 2</u>: Use options as a hedge to offset the downside risk in your portfolio.

Options are available on all eleven Select Sector SPDRs. This feature offers investors an additional tool for managing risk in their portfolios.

5.3 Insurance from portfolio

What product could guarantee a minimum value for financial portfolios, even under adverse market conditions?

Inspired by Black and Merton's work on option pricing (1973), Leland (1985) came up with the idea of adding put options to a portfolio of securities in order to guarantee a minimum value under all circumstances.

Options can be replicated from a risk-free asset and equities through appropriate dynamic positions. In 1976, Leland teamed up with Rubinstein to introduce this technique under the name of portfolio insurance. Subsequently, a number of new management methods were developed, known as portfolio insurance strategies.

5.3.1   Definition

Portfolio insurance strategies are financial strategies whose aim is to limit losses when the market falls, while profiting from the market when it rises.

Three major portfolio insurance strategies are represented here: Option Based Portfolio Insurance (OBPI), which uses options, the Stop-Loss method and Constant Proportion

Portfolio Insurance (CPPI).

5.3.2 Strategy using options

Let's consider a portfolio consisting of a shareSand a European put option on the share
S with maturity T and exercise price K. The value of this portfolio at date T is equal to :

$$V_T = S_T + (K - S_T)^+ = \begin{cases} S_T \text{ si } S_T > K \\ K \text{ si } S_T \text{ } \kappa \end{cases}$$

The price of this option can be interpreted as the premium to be paid to ensure that the
value of the portfolio does not fall below the K floor.

The challenges of this strategy :

- The options available on the market are usually American, and therefore more
expensive than European options.
- The characteristics of the options available on the market are often not in line with
those of the options available on the market.
investor's objectives.
- Regulatory constraints may limit the number of options.

Solution:

Faced with these difficulties, Leland proposed using a monetary asset and the underlying
to synthesize the put.

5.3.3 Stop- Loss strategies

This is the simplest method of portfolio insurance. We will illustrate this method with the
following simple example:

Let x be the initial wealth of an investor.

The amount x is divided between a risk-free asset S_0 and a risky asset S.

V(t) is the value of this portfolio at date t. The stop-loss method consists of

- liquidate all S to buy , as soon as $S_0(t) > S(t)$
- sell all S_0 to buy S, as soon as $S_0(t) < S(t)$

This guarantees that $V(t) \geq S$.

The challenges of this strategy :

Implementing a strategy entails prohibitive transaction costs.

Solution:

We could set an interval around s_0 and only change the composition of our portfolio when S falls outside this interval.

5.3.4   Multiple cushion methods

An investor divides his initial wealth $x > 0$ between a risk-free asset s_0 and a risky asset S.

- The investor chooses a floor value P for the value $V(t)$ of his portfolio.
The difference is called the cushion: $C(t) := V(t) - P$.

- It then sets the amount to be allocated to the risky asset. In this method amount is chosen in the form $m * c_0$ with $m \geq 1$, the factor m is called the multiple.
- Let δ denote the time step and assume that the investor updates his portfolio at each time δk, $k \geq 0$. It is assumed that the investor reallocates his portfolio so that the amount invested in the risky asset at each date δk is equal to $mC\delta k$.
We can then verify that at each date $t_k = \delta k$, the value of the portfolio verifies :

$$V(t_k) - P_k = (x - P) \prod_{i=1}^{k} (1 + m (R_i - r))$$

BIBLIOGRAPHY

Research articles and documents :

Impact of Macroeconomic Announcements on US Equity Prices: 2009-2013. Daniel Nadler, Anatoly B. Schmidt.

Realized and anticipated Macroeconomic conditions forecast stock returns. Alessandro Beber, Michael W. Brandt, Maurizio Luisi. November 2014

Predicting stock returns using industry-relative firm characteristics. Clifford S. Asness, R. Burt Porter, Ross L. Stevens. February 2000.

Sortino, F. and Price, L. Performance Measurement in a Downside Risk Framework. Journalof Investing. Fall 1994, 59-65.

Plantinga, A., Van der Meer, R. and Sortino, F. The Impact of Downside Risk on Risk-Adjusted Performance of Mutual Funds in the Euronext Markets. July 19, 2001.

A. Keating, J. and Shadwick, W.F. The Omega Function. Working paper. Finance Development Center, London 2002.

K., Hossein, T. Schneeweis, and B.Gupta. "Omega as a Performance Measure." June 2003.

Books and courses:

Carl Bacon, Practical portfolio performance measurement and attribution, second edition 2008 p.97-98.

Emmanuel Lépinette, Portfolio Management Course Notes, Université Paris Dauphine, 2011-2012.

Imen Ben Taher, Portfolio Management, Université Paris Dauphine, 2010-2011.

Actuarial thesis :

Mingqian LI and Jianyu CHEN, Historical VaR and Expected Shortfall , Jul 8, 2013.

S. Benseghir, VaR calculation using the historical approach and extreme value theory on a Dexia Asset Management hedge fund, class of 2006.

Maroua Chikhaoui, Portfolio risk management: VaR and CVaR estimation, Maroua Chikhaoui.

Website: www.sectorspdr.com www.quandl.com www.lafinancepourtous.com www.boursedirect.fr

www.evestment.com

www.ressources-actuarielles.net

CONCLUSION

The general objective of financial institutions is to optimize profit and shareholder value by offering a variety of financial services while minimizing the risks incurred. We can deduce from the bibliographical study carried out during our internship that there is no miracle solution to protect against the risks incurred. Risk management is the process of identifying, assessing and prioritizing the risks associated with an organization's activities, whatever their nature or origin. It enables potential risks to be dealt with methodically, in a coordinated and cost-effective way, in order to reduce and control the probability of feared events, and avoid their potential impact.

In our study of SPDR Sector ETFs, we found that, although the Financials (XLF) and Financial Services (XLFS) sectors are among the best performing sectors in terms of average quarterly and annual returns, they also have the best performing Free Cash Flow Per Share ratios, but the highest Volatility Skewness values and the lowest VaR and Expected Shortfall values. This discrepancy can be attributed to the global financial crisis that began in 2007, which amplified the movement and caused stock market prices to plummet, especially for banks and financial institutions.

Our intra-industry and inter-industry variables further explain the cross-asset returns of the equity universe in relation to risk proxies in the larger form of the common market. As we mentioned in Chapter 2, there may be several advantages to decomposition within the industry, but our component analysis shows no advantage over the standard approach. The variables we examined may have explanatory power unrelated to industry classification. If a variable is a price independent of industry, subtracting it means losing information.

In line with the results of the article studied, we found for the annual and half-yearly Free Cash Flow Per Share ratios in Most-Recent Reported View (MR) and As Reported View (AR) dimensions that the Within-industry slope coefficient is slightly greater than the Market-wide slope coefficient, and that the difference between Within-industry slope and Across-industry slope is statistically significant. The Price-to-Book ratio values in dimensions (MR and AR) found show that the Within-industry and Market-wide slope coefficients are almost equal. The observed decrease in power is driven by a strong effect of the across-industry slope measure.

Printed by Books on Demand GmbH, Norderstedt / Germany